Jinfa Cai · Ida A.C. Mok
Vijay Reddy · Kaye Stacey

International Comparative Studies in Mathematics

Lessons for Improving Students' Learning

Jinfa Cai (Chair)
Department of Mathematical Sciences
University of Delaware
Newark, DE
USA

Ida A.C. Mok
Faculty of Education
University of Hong Kong
Pokfulam
Hong Kong, S.A.R.

Vijay Reddy
Human Sciences Research Council
Durban
South Africa

Kaye Stacey
Melbourne Graduate School of Education
University of Melbourne
Melbourne, VIC
Australia

DOI 10.1007/978-3-319-42414-9

Library of Congress Control Number: 2016946614

Acknowledgements

We are grateful for the assistance of reviewing and formatting this paper by Victoria Robison and Sylvia Hannan.

Contents

International Comparative Studies in Mathematics: Lessons for Improving Students' Learning

1 Introduction

Comparing is one of the most basic intellectual activities. We consciously make comparisons to understand where we stand, both in relation to others as well as to our own past experiences. There has been a long history of international comparative studies in education (Alexander 2000). Especially in the past several decades, many international comparative studies of mathematics have been conducted, either to examine differences in mathematical proficiency and dispositions among students from various countries or to understand the possible influence on the observed differences of various factors such as curriculum, teacher preparation, quality of classroom instruction, and parental involvement. Some of these studies are large-scale, and others are small-scale in-depth analyses from cognitive or social perspectives. These international comparative studies in mathematics are valuable because they provide a large body of knowledge showing how students do mathematics in the context of the world's varied educational institutions. In addition, they examine the cultural and educational factors that influence the learning of mathematics.

Comparative studies provide information on students' achievement as examined in the context of the world's varied educational institutions, and they also help identify effective aspects of educational practice. Postlethwaite (1988) identified four objectives of comparative studies:

- Identifying what is happening in different countries that might help improve education systems and outcomes;
- Describing similarities and differences in educational phenomena between systems of education and interpreting why these exist;
- Estimating the relative effects of variables that are thought to be determinants of educational outcomes (both within and between systems of education); and
- Identifying general principles concerning educational effects.

J. Cai et al., *International Comparative Studies in Mathematics*,
ICME-13 Topical Surveys, DOI 10.1007/978-3-319-42414-9_1

The field of mathematics education also sees international comparative studies as a means to generate theories and to identify practices that may improve students' learning. According to Cai (1995), international comparative studies in the domain of mathematics provide mathematics educators with opportunities to identify effective ways of teaching and learning mathematics. Examination of what is happening in the learning of mathematics in other countries helps researchers and educators understand how mathematics is taught by teachers and how it is learned and performed by students in different countries. It also helps them reflect on theories and practices of teaching and learning mathematics in their own culture. Stigler et al. (2000), themselves researchers conducting international studies, explain the value of this research on trends over time and context in a more nuanced way:

> Consequently, we may be blind to some of the most significant features that characterize teaching in our own culture because we take them for granted as the way things are and ought to be. Cross-cultural comparison is a powerful way to unveil unnoticed but ubiquitous practices (pp. 86–87).

1.1 How Are Large Scale Studies Used?

Large-scale international comparative studies have proved useful in a wide variety of ways to improve mathematics learning. For example, the book by Stacey and Turner (2015a) includes a section on the impact in 14 countries of the Programme for International Student Assessment (PISA) conducted by the Organisation for Economic Co-operation and Development (OECD). The reports of local mathematics educators were that every country looks at its ranking overall and its position with respect to countries that are similar in important respects. Increasingly, countries use PISA to monitor progress and measure the success of their own initiatives to improve student learning by comparing results from cycle to cycle. Nine of the 14 contributions indicated that the Framework of the PISA assessment (OECD 2013a) had influenced directions for curriculum reform. PISA's emphasis on mathematical literacy (briefly, the ability to use mathematical knowledge; but see OECD 2013a, p. 25) is a major influence. PISA focuses on the uses of mathematics (from simple to complex), because this utilitarian aspect of mathematics is the key to its economic importance and to the well-being and life chances of each citizen. Sometimes the new directions have been to increase emphasis on a mathematical content category (e.g., Uncertainty and Data), but more commonly new directions promote emphasis on mathematical literacy as a goal and/or give attention to the processes and the fundamental mathematical capabilities for using mathematics. These changes in direction were reported in work with national standards and curriculum materials as well as direct work with teachers and students. Three contributions in Stacey and Turner (2015a) demonstrated how PISA items and the framework could be used with teachers to expand conceptions of worthwhile mathematics. Four contributions described improvements to national assessment practice, some of them technical, but some also assessing broader goals of mathematics education. Finally, four contributions highlighted how PISA's identification of inequalities in student achievement is leading to action on equity (gender, geographical, socioeconomic, etc.). Major teacher education initiatives created in response to PISA results were reported from two countries. Reading these contributions demonstrates the wide range of impacts that the PISA surveys have had, and also the context-sensitive ways in which mathematics educators have adapted the PISA frameworks and interpreted them for their local situations.

An early study (Elley 2002) of countries whose participation had been supported by the World Bank reported similar results on the use of data from TIMSS-R from the International Association of the Evaluation of Educational Achievement (IEA). TIMSS provided education authorities with reliable data on education systems, often for the first time, and provided excellent training for the national teams on modern assessment practice. The achievement scores were the most important findings, both absolute and relative to similar countries, followed by trends over time. TIMSS results motivated some changes in policy and a lot of work revising standards, curriculum, and teacher resource materials.

1.2 Unique Findings from Small Scale Studies

Contrasting in scale to TIMSS and PISA, there are a number of relatively small-scale international comparative studies (e.g., Cai 1995, 2005; Cai et al. 2014; Ma 1999; Silver et al. 1995; Song and Ginsburg 1987; Stevenson et al. 1990; Stigler et al. 1990). Over 20 years ago, Bradburn and Gilford (1990) suggested that for small in-depth cross-national studies, relatively small and localized samples in a small number of sites are acceptable. Perhaps the assumption is that the differences arising from national cultures are so fundamental that they will be evident in data from any sample. The information from these smaller studies can play an important role in educational research and policy development. They can reveal unique findings beyond the scope of the large-scale studies and also complement the large-scale studies by providing deep understanding about different societies and education, thereby enhancing interpretations and implications.

1.3 What Is an International Comparative Study?

In this section we clarify the terms used in this paper, and incidentally highlight the range of studies that might be called international comparative studies. Several different terms are used in the literature, such as international comparative studies, cross-national studies, and cross-cultural studies. Possibly, international means multiple countries and cross-national means two or more different countries. In this panel, we use the phrase 'international comparative studies' to refer to all those involving at least two "countries," with an intention to compare at the country level. "Countries" should be interpret in a broader sense. We include studies that are small and large, qualitative and quantitative, and initiatives of government or individual researchers.

With this definition, we see international comparative studies in mathematics evolving from informal observations to the examination of performance differences, and from the examination of contributing factors to performance differences to the generation of theories, actions, and policies based on international comparative studies. In terms of scale, international comparative studies range from small-scale studies involving a few classes from two countries to the large-scale studies like TEDS (M), TIMSS, and PISA, with upwards of half a million participants.

Globalization and technological progress have created far more opportunities for international comparative studies. Thus, international comparative studies reflect changes in the world. Accessible air travel enabled visits; video enabled transporting classroom behavior to researchers around the world; computers gave the ability to deal with large data sets; email enabled timely interchange on international projects. Globalization, with both co-operation and competition between nations, facilitates an interest by scholars and also by governments in what others are doing. Personal links between far-apart researchers give rise to smaller projects.

International comparative studies not only intend to compare different school systems and different traditions and cultures of schooling, but also to create awareness of different possibilities for teaching mathematics and improving students' learning of mathematics.

1.4 What Lessons Can We Learn from International Comparative Studies?

The focus of this ICME-13 Topical Survey is to discuss the ways to use international comparative studies to improve students' learning. We take a strong position that the main purpose of educational research is to improve student learning. In fact, according to the OECD, the group that administers PISA, test results are intended to show where countries stand, and motivate policymakers to identify shortcomings and suggest potential strategies for reform.

The highest profile international comparative studies, such as PISA and TIMSS, have had a significant impact on thinking about education around the world, especially related to the broad characteristics of educational systems and government policy, where mathematics is just one of several important components. The fundamental purpose of large-scale studies like PISA and TIMSS is to meet governments' need for objective evidence to monitor educational outcomes, to demonstrate possibilities, and to assist in developing new policies. There is no sign of the international comparative studies slowing down, and the purpose of this paper is not to advocate putting a break on them. Instead, it is to take a step back and reflect on the studies and the lessons we can learn from these studies.

In this Topical Survey, we discuss four of the many lessons we can learn from international comparative studies for improving students' learning, and suggest future directions. The four lessons, which are chosen to represent different styles and strands of the work, are related to (1) understanding students' thinking, (2) examining the dispositions and experiences of mathematically literate students, (3) changing classroom instruction, and (4) making global assessment research locally meaningful. We decided to focus on these four aspects because of their importance for the impact on students' learning. The first two lessons focus on students' mathematical thinking and achievement. The third lesson focuses on classroom instruction, and the fourth lesson focuses on policy in the local context. We have used both small- and large-scale international comparative studies to illustrate the lessons we can learn.

2 Survey on State-of-the-Art

2.1 Lesson 1: Understanding Students' Thinking

2.1.1 Findings from Two Examples

To illustrate the lessons learned in this aspect, let's first look at two examples. The first example is related to the Pizza Ratio Problem in Fig. 1 and the second example in Figs. 3 and 4 includes two arithmetic averaging problems.

****

Here are some children and pizzas. 7 girls share 2 pizzas equally and 3 boys share 1 pizza equally.

A. Does each girl get the same amount as each boy?
Explain or show how you found your answer.

B. If each girl does not get the same amount as each boy, who gets more?
Explain or show how you found your answer.

****

Fig. 1 The Pizza Ratio Problem

The Pizza Ratio Problem, as seen in Fig. 1, requires students to justify whether each girl gets the same amount of pizza as each boy, and if not, who gets more. Cai (2000) identified eight convincing arguments Chinese and U.S. students have used in their solutions to this problem (see Fig. 2).

Of those providing complete and convincing arguments, the majority (90 %) of the Chinese students provided a numerical argument (Argument 1). Only a few of the Chinese students used any of the other types of argument. In their numerical arguments, Chinese students tended to use fractions with a common denominator

****

Convincing Argument 1

Each boy will get 1/3 of a pizza and each girl will get 2/7 of a pizza. If you compared 1/3 with 2/7, you would know that 1/3 is bigger than 2/7 by transforming them to have common denominators (1/3 = 7/21 and 2/7 = 6/21 and then 7/21 - 6/21 = 1/21) or decimals (1/3 =0 .33 and 2/7 = 0.29 and then 0.33 - 0.29 = 0.04).

Convincing Argument 2

If there were six girls, each girl and each boy would have the same. But you have 7 girls, so each girl gets less than each boy.

Convincing Argument 3

Three girls share one pizza, and another three girls share another pizza. Each of these six girls will get the same amount of the pizza as each of the three boys. But one of the girls has no pizza. So, each boy will get more.

Convincing Argument 4

Three girls share one pizza and the remaining four share one pizza. Each piece that each of the remaining four girls gets are smaller than those the boys get. So the boys get more.

Convincing Argument 5

7 girls get two pizzas, and 3 boys get one pizza. Girls have twice as much pizza as boys. But the number of girls is more than twice the number of boys. So the boys get more.

Convincing Argument 6

Each pizza was cut into 4 pieces. Each girl gets 1 piece and then there is 1 piece left over. Each boy gets 1 piece and then there is 1 piece left over. 1 piece left over must be shared by the 7 girls, but the 1 piece left over will be shared by three boys. So the boys get more.

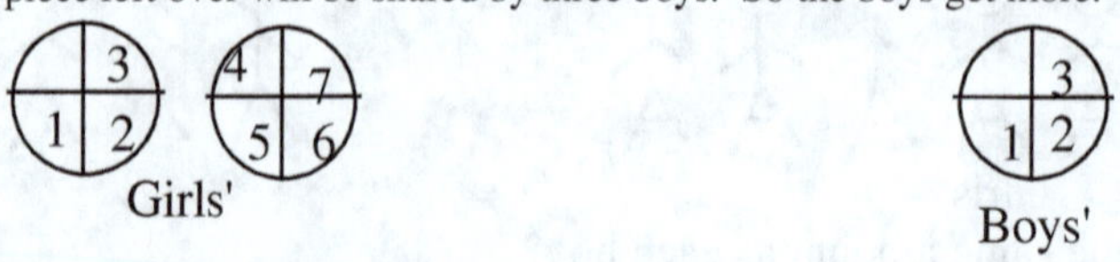

Convincing Argument 7

7/2 = 3.5 and 3/1 = 3. Therefore 3.5 girls will share one pizza and 3 boys will share one pizza. Thus, each boy gets more.

Convincing Argument 8

Each pizza is cut into 21 pieces. Each girl will get 6 pieces and each boy will get 7 pieces.

****

Fig. 2 Eight arguments students used in their solutions to the Pizza Ratio Problem

instead of decimals. In fact, five times as many Chinese students used fractions as decimals in their numerical arguments. Argument 4 was the argument most commonly used by the U.S. students (29 %). About one-fifth of the U.S. students used Argument 1, and one student used Argument 8. The remaining arguments were each used by about 10 % of the U.S. students. Examining the spread of these eight arguments is quite informative in revealing what students think about this problem, but if we quantitatively score these responses on correctness alone, they are all the same.

The second example is related to a pair of arithmetic averaging problems, shown below in Figs. 3 and 4. The first task required students to find the average of four numbers. The second task required students to find the missing number in a pictograph that showed the first three numbers and gave the average of all four numbers. A correct solution to this second task requires a well-developed understanding of the concept of averaging. In both tasks, students were asked to provide a numerical answer and to explain how they found their answer.

****

The Can Averaging Problem (Task 1)

For their club's food drive, Tasha has 11 cans, David has 6 cans, Jeffrey has 5 cans, and Dwayne has 2 cans. What is the average number of cans for those four people? Explain how you found your answer.

****

Fig. 3 The first averaging task

****

The Hats Averaging Problem (Task 2)

Angela is selling hats for the Mathematics Club. This picture shows the number of hats Angela sold during the first three weeks.

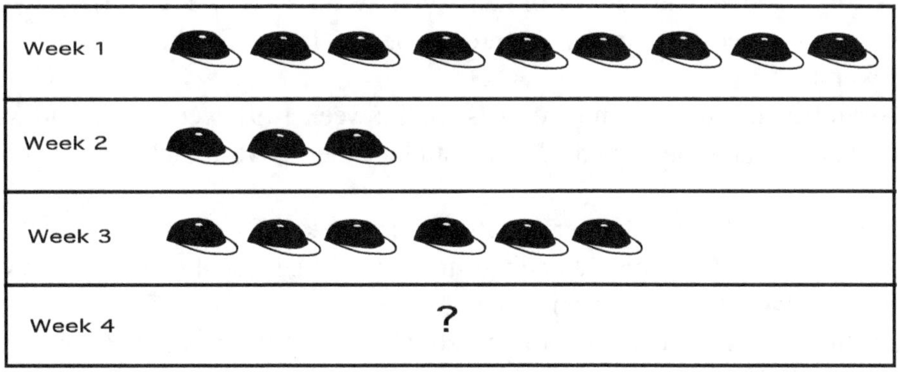

How many hats must Angela sell in Week 4 so that the average number of hats sold is 7? Show how you found your answer.

****

Fig. 4 The second averaging task

Table 1 Correctness of numerical answers for both averaging problems

Task 1	Task 2			
	Correct		Incorrect	
	U.S.	China	U.S.	China
Correct	42 %	67 %	24 %	23 %
Incorrect	3 %	4 %	31 %	6 %

Table 1 shows the percentages of both U.S. and Chinese students who solved each of the two average problems. A larger percentage of Chinese (67 %) than U.S. students (42 %) provided correct answers for both tasks. In addition, a larger percentage of U.S. (31 %) than Chinese students (6 %) incorrectly answered both tasks. Interestingly, about one-fourth of the U.S. students and one-fourth of the Chinese students had the correct answer for the first task, but not for the second one. That is, about a quarter of these U.S. students and Chinese students knew the averaging algorithm, but lacked the conceptual understanding of averaging necessary to solve the problem. About 5 % of the U.S. students and about 5 % of the Chinese students got incorrect answers for the first task, but a correct answer for the second task. This appeared to be due to calculation errors in solving the first task. In summary, the results from this example show that about 25 % of the Chinese and U.S. students knew the averaging algorithm, but did not necessarily conceptually understand averaging sufficiently robustly that they were able to determine when and how to "add and divide."

For both samples, the most common error type on the second problem was an "incorrect use of the computational algorithm." Six error types are identified and shown below. These errors suggest that students know the procedure to add and divide to find the average, but do not understand the concept.

1. The student added the number of hats sold in week 1 (9), week 2 (3), and week 3 (6), then divided the sum by 3, and got 6. However, the average was 7. Therefore, the student added 3 to the sum of the numbers of hats sold in the first 3 weeks, then divided it by 3, and got 7, and then gave the answer 3.
2. The student added the number of hats sold in week 1 (9), week 2 (3), and week 3 (6), then divided the sum by 3, and got 6, $6 + 1 = 7$. So the student gave the answer 1.
3. The student added the number of hats sold in week 1 (9), week 2 (3), and week 3 (6), then divided the sum by 3. The student then gave the quotient (6) as the answer.
4. The student added the number of hats sold in week 1 (9), week 2 (3), week 3 (6), and the average (7), then divided the sum by 4. The student then gave the whole number quotient (6) as the answer.
5. The student added the number of hats sold in week 1 (9), week 2 (3), and week 3 (6), then divided the sum by 4. The student then gave the quotient (4.5) as the answer.
6. The student added the number of hats sold in week 1 (9), week 2 (3), and week 3 (6), then divided the sum by 7. The student then gave the whole number quotient (2) as the answer.

2.1.2 Needs and Benefits of In-depth Studies to Understand Students' Thinking

In international comparative studies, especially in large-scale studies, results are usually reported quantitatively, by assigning numerical scores. The quantitative analysis is useful for providing an overall picture of students' performance in mathematics and it enables the examination of patterns and relationships among variables and the opportunity to test them statistically. In particular, quantitative analysis allows for examining variables which may predict students' learning outcomes. However, scoring on the basis of correctness alone will conceal some important aspects of students' performance. The results from both the Pizza Ratio Problem and the Averaging Problems have demonstrated that different students can use different strategies to obtain a particular score level. Similarly, students in different places may exhibit different mathematical misconceptions at a particular score level. Such important differences in students' mathematical thinking may reflect the differences in teachers' beliefs and instructional practices (e.g., Cai 2004; Cai and Ding 2015; Cai et al. 2014; Cai and Wang 2010).

Given the fact that simple comparisons of international rankings provide little guidance for understanding and improving students' mathematics learning, it is important to understand international performance differences by other means. The in-depth, small-scale international comparative studies can provide unique opportunities for us to understand students' mathematical thinking. The more information teachers have about what students know and think, the more opportunities they can create for student success. Teachers' knowledge of students' thinking has a substantial impact on their classroom instruction and, hence, upon students' learning. Therefore, in order to provide the education community with a deeper understanding of the teaching and learning of mathematics, it is essential for some international comparative studies to provide evidence of students' thinking and reasoning beyond correctness of answers to mathematical problems. The evidence of students' thinking includes the qualitative analysis of solution strategies, mathematical errors, mathematical justifications, and representations (Cai 1995).

2.1.3 Nurturing Creativity and Critical Thinking Skills

The world has been changing dramatically, and these changes are happening much faster than we anticipated. Today, possessing a large amount of knowledge and information is not sufficient. Instead, in this continually changing world, the most important qualities we can help our students develop are the abilities to think independently and critically, to learn, and to be creative. In his best-selling book, *The World Is Flat*, Friedman (2005) pointed out that "there may be a limit to the number of good factory jobs in the world, but there is no limit to the number of idea-generated jobs in the world" (p. 230). K-16 education in general, and mathematics and science education in particular, has a responsibility for nurturing students' creativity and critical thinking skills.

A conventional strategy is one that is usually taught in the classroom; in contrast, a non-conventional strategy may not be taught in the classroom and may evolve from the students' novel explorations. The results for the Pizza Ratio Problem above showed that Chinese students were more likely to use the conventional strategies of Argument 1 above, comparing fractions with common denominators or comparing decimals. However, only about 20 % of the U.S. students used this conventional strategy. In contrast, the vast majority of the U.S. students used somewhat non-conventional strategies. This example shows a dilemma we face. Clearly, the conventional strategy is quite efficient and it can be easily applied to solve other similar problems, but this conventional strategy shows little originality. Whilst non-conventional strategies show the originality of students' thinking, they are also task-specific and less easily applied to solving other problems, especially those that involve bigger numbers. The results from this particular example may suggest the effectiveness of Chinese classroom instruction in developing students' efficient strategies and the effectiveness of U.S. classroom instruction in developing original mathematical thinking. Ideally, we would hope that instruction can foster students' learning of efficient problem solving strategies and develop original mathematical thinking. If that is one of the goals for school mathematics, we have to seriously investigate the classroom instruction in both nations so that each can learn from the other.

We can extend the discussion of the strategies to solve the Pizza Ratio Problem to a broader context. It is possible that educators and government officials tend to believe that the USA does a better job of nurturing students' creativity than Asian countries do. For example, India's Prime Minister, Manmohan Singh, revealed that two-thirds of the nation's universities and 90 % of its degree-granting colleges were rated as below average, and that university curricula were typically not in alignment with the needs of employers or job seekers (Bharucha 2008). On the other hand, in the USA, several recent reports call for the U.S. to learn from Asia because it is believed that Asian countries like China, India, and Singapore are much more effective in mathematics and science education, thus posing a major threat to the global competitiveness of the USA (Asian Society 2006). However, some Asian-born scholars believe that Asian countries should learn from the USA about science, technology, engineering, and mathematics (STEM) education (e.g., Bharucha 2008; Zhao 2008) because the U.S. does a better job of nurturing creativity.

2.1.4 Calling for More In-depth, Small-Scale International Comparative Studies

Because of the obvious advantages of international comparative studies such as PISA and TIMSS, such large-scale studies continue to dominate the public and scholarly discourse about mathematics education worldwide. We should not underestimate the value of in-depth, small-scale international comparative studies to understand students' mathematical thinking. Such studies can complement the large-scale international comparative studies to provide insight into students' mathematical thinking and reasoning. Therefore, more effort is needed to conduct such small-scale comparative studies.

The in-depth, small-scale international comparative studies can also start to explore many urgent and important research questions. For example, is there really a creativity gap between students in Asian countries and the USA? How can we best assess students' creativity in general and creativity in STEM in particular? How should creativity be nurtured? Future international comparative studies and international collaboration should answer these questions empirically.

There are distinct advantages for individual researchers to collaborate on in-depth, small-scale international comparative studies on students' thinking, because fewer resources are required. The recent rapid increase of international comparative studies on curriculum is a good example (Lloyd et al. 2016). Many individual researchers chose to focus on certain aspects of curriculum and they conducted comparative analysis across nations. These studies provided new insights into the content and design of mathematics textbooks and generated key questions about relationships between written curricular materials and students' opportunities to learn. Similar to curriculum studies, we hope that individual researchers can also focus on certain aspects of mathematics to conduct in-depth, small-scale international comparative studies on students' thinking.

2.2 Lesson 2: Examining the Dispositions and Experiences of Mathematically Literate Students[1]

As noted above, the results of large-scale studies have many lessons for educational systems related to overall achievement and links to student background variables, but what lessons can they offer to mathematics educators to inform more detailed questions about teaching mathematics? One specific aspect of PISA 2012 that is of direct interest to mathematics educators and that provides data of a different character than that which mathematics educators might associate with large scale studies relates to the teaching of mathematical literacy. Mathematical literacy, the achievement construct that is measured by PISA, is the ability to use mathematical knowledge in situations that are likely to arise in the lives and work of citizens in the modern world. A precise definition is given by the OECD (2013a, p. 25).

Before 2012, PISA surveys reported overall scores for mathematics achievement for countries and nominated subgroups of students (by gender, socioeconomic status, etc.) with sub-scores for achievement by four content categories (e.g., Change and Relationships) and three 'competency clusters.' In 2012 when

[1]Note that PISA 2012 terminology is used throughout this document. Both countries and economies participate in PISA; in this document the word 'country' refers to both. Kaye Stacey was Chair of the Mathematics Expert Group for PISA 2012. The opinions expressed here are her own.

mathematics was the major domain, the PISA survey was able to examine many more aspects of its key concept of mathematical literacy. Hence it investigated: the achievement profiles of students across the processes that are involved in exercising mathematical literacy; the learning opportunities that contribute to achievement and how they vary across countries; in-class experiences and dispositions that influence mathematical literacy; and the effect of classroom experiences with mathematical literacy on more general students' attitudes. This section briefly outlines some of the lessons that can be drawn from this work, and in so doing may draw attention to possibilities for some new directions and new research questions for mathematics educators. It is clear that this work is just at the beginning.

2.2.1 Country Profiles of the Processes of Mathematical Literacy

Using mathematics to meet a real-world challenge involves three 'processes' depicted in Fig. 5.

- Formulating situations mathematically (abbreviated to Formulate),
- Employing mathematical concepts, facts, procedures, and reasoning (Employ), and
- Interpreting, applying, and evaluating mathematical outcomes (Interpret).

Readers will note the intentional similarity of Fig. 5 to many familiar diagrams depicting the mathematical modeling cycle. The Formulate process transforms the real world challenge into mathematical form by identifying variables and relationships and making assumptions. The Employ process takes place largely within the mathematical world, using the mathematical knowledge and skills that form the bulk of school education. The Interpret process (which for PISA purposes includes both interpretation and evaluation of the real world solution) transforms the mathematical answers back to the real world context and judges their real world adequacy.

PISA 2012 was able to measure the performance of students on each of these three processes separately, and this revealed interesting country patterns and

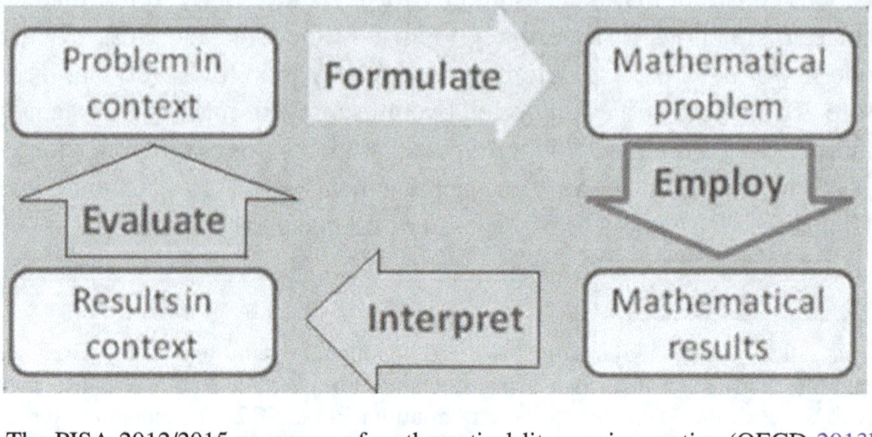

Fig. 5 The PISA 2012/2015 processes of mathematical literacy in practice (OECD 2013b)

differences for the first time. The average score for overall mathematical literacy in PISA 2012 across the OECD was 494 made up of 492 for Formulate, 493 for Employ, and 497 for Interpret. In other words, Formulate items were found to be more difficult than the average item, and Interpret items were easier. This is despite the intention of the survey design to select items to measure each process so that the three overall means would be the same. Most international studies show that boys have mathematics achievement than girls: in PISA 2012, the biggest difference between boys and girls was in Formulate.

Where was the strength of the top performing countries? Perhaps surprisingly, 9 of the 10 top performing countries had their own highest scores in Formulate. Top performers were generally Asian countries, which stereotypes might have predicted would instead have their greatest strength in calculation and hence in the Employ process, rather than in Formulate. Interestingly, the four highest performing countries had their own lowest score on Interpret—the set of items that were easiest for most students internationally. Earlier small-scale studies (Cai 1995; Cai and Silver 1995) reported similar findings regarding the solution of a 'division with remainder' problem. Chinese students outperformed U.S. students in the computational phase of solving the division problem involving a remainder, but students in both samples had similar cognitive difficulties in making sense of the answer, i.e. in PISA's interpretation process.

Other countries showed different patterns. The Netherlands, Denmark, and Sweden had their higher scores on both Formulate and Interpret (the two processes where real-world contexts matter), and their lowest on the intra-mathematical Employ. Non-Asian English-speaking countries (Canada, Australia, New Zealand, UK, USA) were relatively stronger in Interpret only. Nine European countries were relatively low in Formulate but higher in both Employ and Interpret. All of these results warrant further investigation, both to understand country patterns better (and their possible links to curriculum and teaching practices) and to examine how the items have functioned in different contexts. A start has been made by Stacey and Turner (2015b). Better understanding of results such as these requires both large scale and small scale research, looking at standards, curriculum, teaching, learning and assessment.

Large scale studies rely on scoring that is highly specified, reliable and efficient, but the section above illustrates that the item set can be structured so that the quantitative results can provide considerably more information about students' proficiency than just one single score.

2.2.2 What Mathematics Experiences Build Mathematical Literacy?

Since PISA's construct of mathematical literacy involves mathematics that is likely to be useful to citizens in all walks of life, it is of interest to know whether a curriculum produces better mathematical literacy outcomes if it is oriented towards abstract mathematics or to its applications. At one extreme, it might be hypothesized, for example, that students need deep exposure to applications of mathematics in order to do well, but at the other extreme, students may be able to apply mathematics if they are taught abstract mathematics only, because they know the principles to apply.

To answer this question, some of the students who took part in the PISA 2012 study were asked to rate how confident they felt about solving a set of mathematics tasks, and later in the questionnaire, were also asked to rate how frequently they had encountered similar tasks in class. These student reports were used to develop an index of exposure to applied mathematics, and an index for formal mathematics[2] (OECD 2013b), thereby measuring opportunity to learn these two types of mathematics. The sample tasks included 'formal' (pure) mathematics items without any context, such as solving a linear equation or finding the volume of a box, and 'applied' mathematics items such as calculating times using a train timetable, calculating petrol consumption, and interpreting a misleading graph in a newspaper (a released PISA item). Full details of the measures and precise results are given in the official reports of PISA (OECD 2013b, c) and the general summaries below draw heavily on these chapters.

The findings from 2012 indicated that performance in PISA is very strongly related to opportunities to learn formal mathematics and secondarily to opportunities to learn applied mathematics. The relationship of PISA performance with exposure to formal mathematics is positive and linear at the student, school, and country level. For applied mathematics, all three relationships are quadratic. The more frequently students are exposed to applied mathematics problems, the better is their PISA performance, but only up to a point—high exposure is associated with a decline in performance. One explanation for the curvilinear relationship may be that low-performing students are placed in classes or schools with a focus on the 'everyday' applications of mathematics and high-performing students are taught more formal mathematics. PISA data reveals the relationship but it cannot provide a causal explanation. To explain the finding, more detailed knowledge of individual school systems and teaching practices is needed. Only with this information can we understand the significance of the findings.

Japan and the Netherlands, both high-achieving countries, provide interesting contrasts. Students in Japan (mean PISA score 536, Formulate 554, Employ 530, Interpret 531) reported low exposure to applied mathematics and high exposure to formal mathematics (OECD 2013b, c), whereas students in the Netherlands (mean PISA score 523, Formulate 527, Employ 518, Interpret 526) reported high exposure to 'applied mathematics' and low exposure to 'formal mathematics,' perhaps indicating the influence of Realistic Mathematics Education (RME) there. The Japanese students perceive an emphasis on formal mathematics, but through their total experience they have learned to identify mathematical relationships within real situations and to create appropriate models. How this has happened is a question for further investigation.

---

[2]The correct name is "index of experience with pure mathematics," rather than formal mathematics. It is used here to bring slightly different concepts together in this broad summary.

The PISA surveys and many other studies frequently find that there is a large variation in the extent to which students attending the same school report exposure to different teaching strategies and teacher behaviors. This is called the within-school variation. PISA 2012 found that there is also large within-school variation in students' reports of exposure to formal and applied mathematics tasks. When the variability within each OECD country is divided between within-school and between-school factors, on average the between-schools factors account for only 3 % of the overall variation in students' reported experience with applied mathematics tasks and 9 % of the overall variation in student exposure to formal mathematics tasks. It is possible that students' individual experiences within a class and what different members of a class might recall from even the same lesson will account for a lot of the within-school variation. However, it does seem that treatment of formal mathematics is decided somewhat more at the whole-school level (possibly even at the country level) than are the applied situations that draw on the abstract content. As with all other results, the specific statistics for countries and sub-populations will be influenced by country-specific factors, and indeed even the 'real' causes of identical results may be different for different countries. For example, an educational system where heterogeneity in student performance is dealt with by sending students to different types of schools may show comparatively large between-school variation even if the between-school variation for similar types of school is small because of tight control. On the other hand a very loosely controlled educational system with a tradition of teachers working closely together within schools may also exhibit large between-school variation. Allocating students to different types of mathematics classes within each school is a potential source of comparatively large within-school variation, as is a teacher tradition of autonomy.

2.2.3 Students' Disposition Toward Formal and Applied Mathematics

PISA 2012 also provided some important lessons about student dispositions. Dispositions are especially relevant in the international governmental climate where the importance of mathematical literacy to economic well-being is widely acknowledged, and many countries seek to increase participation in STEM careers. This necessitates attention to students' dispositions. Students' responses to questions about how confident they are in solving the sample tasks were used to create an index of mathematics self-efficacy (confidence). Some patterns in the results provide important lessons for mathematics educators. As would be expected, there was a high association between students' reporting that they had encountered a task frequently and their reported confidence in solving it. For example, 74 % of OECD students who reported having frequently encountered the task about using a scale on a map reported feeling confident about solving the task, whereas only 32 % of those who reported never having encountered it felt confident (OECD 2013c). With the formal mathematics problems, almost all OECD students who reported having frequently encountered formal mathematics tasks felt confident about solving such tasks. However, confidence in solving applied mathematics problems was much lower even when students reported having frequently encountered such problems. One possible explanation is that solving applied mathematics problems requires both a good understanding of an underlying abstract structure as well the ability to analyze the real situation—in other words it requires the PISA mathematical processes of Formulate and Interpret, as well as Employ. More detailed analyses of the PISA data and further studies seem warranted to understand this potentially important phenomenon. These results suggest that exposure matters for students' mathematics self-efficacy. Moreover, further analysis demonstrated that the relationship between exposure and feelings of self-efficacy is 'narrow' in the sense that frequent exposure to any one of the tasks had very little influence on confidence in solving any of the others, regardless of whether they were formal or applied. One lesson from this is that there may be much more to learn from studying self-efficacy at the task level, in addition to the conventional construct of a global self-efficacy towards mathematics.

There is a well-known gender difference in confidence in mathematics, which is evident in a large majority of countries. PISA 2012 (OECD 2013b) found that this gender difference was not observed with the formal mathematics tasks that match likely classroom content (such as solving a linear equation). However, there were large gender differences with the applied tasks. For example, across OECD countries, 75 % of girls reported being confident or very confident when calculating a 30 % discount on a TV, compared to 84 % of boys. Differences were even greater when the context was stereotypically male. Across OECD countries, 67 % of boys but only 44 % of girls reported feeling confident about calculating the petrol consumption rate of a car. Why is important? The self-efficacy index explains more than 30 % of the variance in proficiency for about a third of the PISA participating countries (OECD 2013b). Hence, these gender gaps evident in self-efficacy for applied mathematics problems are likely to have an impact on gender differences in achievement and career choices.

Dispositions are linked to exposure to formal and applied mathematics. The first report for PISA 2012 (OECD 2013b) described the relationship between students' exposure to formal and applied mathematics problems and student engagement, drive and motivation, and mathematics self-beliefs. Overall, students who reported having been more frequently exposed to formal mathematics tasks reported more positive engagement, drive, motivation, and self-beliefs. The same relationship worked for applied mathematics tasks. Because of the inter-relationships between positive disposition and other variables (e.g., gender, mathematics performance), further analyses were undertaken.

One set of analyses compared students with similar performance in mathematics. At each level of performance, exposure to applied mathematics tasks was strongly and positively associated with students' drive, motivation, and self-beliefs. The association of these characteristics with exposure to formal mathematics tasks was weaker (probably because formal mathematics exposure is more strongly associated with performance) and in many countries was not present. For example, intrinsic motivation to learn mathematics is positively associated with exposure to applied mathematics problems among students of equal performance in nearly all countries and with exposure to formal mathematics problems in a large majority of countries.

The first report of PISA 2012 (OECD 2013b) summarizes these findings as follows:

> Results on the association between students' reported experience with formal and applied mathematics problems, mathematics performance, and engagement, drive, motivation and self-beliefs suggest that students who are frequently exposed to both formal and applied mathematics problems fare particularly well: they perform at higher levels in mathematics and enjoy greater drive, motivation and more positive self-beliefs (p. 134).

However, once again, it is important to acknowledge that the direction of causality cannot be decided from this data and analysis. It is possible, for example, that differences in students' reports of exposure may reflect other variables. Better students may remember more instances of exposure or indeed may have participated in them more frequently. Some teachers might only present applied mathematics problems to students who have mastered relevant formal mathematics. Other teachers might use applied mathematics tasks as a way to spark interest and motivation among lower-achieving students. This is yet another series of links for mathematics educators to study, using a variety of within-country and cross-national research designs, in the search for ways to help all students achieve their full mathematical potential.

The large scale studies provide a wealth of information for mathematics education researchers and policy makers. However, there is a clear need for more engagement from the mathematics education community to contribute to understanding the findings better, and hence to assist us all in learning the right lessons about student learning.

2.3 *Lesson 3: Changing Classroom Instruction*

This section draws upon the work of two international studies of teaching practice, the TIMSS Video Study and the Learner's Perspective Study (LPS), to zoom into what we may learn from international comparative studies focusing on classroom instruction.

2.3.1 Complementary Roles of TIMSS Video Study and Learner's Perspective Study

The first TIMSS Video Study took place in 1995 (Stigler and Hiebert 1999) and studied national samples of eighth grade mathematics lessons from Germany, Japan, and the USA. The conclusion reported in The Teaching Gap (Stigler and Hiebert 1999) was that teaching was a cultural activity. The design of this TIMSS Video Study combined methodologies of qualitative classroom research and large-scale survey research. The study captured a snapshot of eighth grade mathematics lessons of a statistically rigorous national sample of teachers, who were ordinary teachers teaching lessons that they routinely teach. The tradition of large-scale surveys of a national sample made it possible to describe average teaching based on a probabilistic statistical method. Building on the first TIMSS study of student achievement, the TIMSS 1999 Video Study (Mathematics) compared teaching practices in the USA and places that showed higher performances than the USA in TIMSS: Australia, the Czech Republic, Japan, the Netherlands, Switzerland, and Hong Kong. This TIMSS Video Study documented typical teaching practices in these countries. The study aimed to find pictures of what average teaching looked like in different countries. Another aim was that studying lessons from different cultures would give researchers and teachers the opportunity to discover alternative ideas about how we might teach mathematics (Stigler and Hiebert 2004).

The Learner's Perspective Study (Clarke et al. 2006a) was designed to examine the practices of eighth grade mathematics classrooms in Australia, Germany, Japan, and the USA in an integrated, comprehensive way. The project was originally designed to complement other international studies that reported national norms of student achievement and teaching practices. Since its inception, the LPS community has expanded to become a research community consisting of researchers in Australia, China, the Czech Republic, Finland, Germany, Israel, Japan, Korea, New Zealand, Norway, the Philippines, Portugal, Singapore, Slovakia, South Africa, Sweden, the United Kingdom, and the USA. LPS aimed to juxtapose the observable practices of the classroom and meanings attributed to those practices by the teachers and the students. The LPS research design documents sequences of at least 10 lessons, using three video cameras, supplemented by the reconstructive accounts of classroom participants obtained in post-lesson video-stimulated interviews and by test and questionnaire data, and copies of student work. In each participating country, data generation focused on three teachers identified as competent by the local mathematics education community. The rich data set provided researchers with the opportunity to reconstruct complementary accounts of the lessons and analysis via different theoretical frameworks and focuses, giving the opportunity for comparisons of the instruction in mathematics classrooms in different countries. The design of the study did not aim for a representative national sample like the 1995 TIMSS Video Study. Instead, the project provided a way to understand what competent teachers, recognized locally in different cultural settings, might make possible.

2.3.2 Lesson Structures and Lesson Events

While seeking a viable unit of international comparative analysis of mathematics lessons, the LPS research team used the coding in the TIMSS Video Study for patterns of lesson structures—reviewing, demonstration of the problem for the day, practicing and correcting seatwork, and assigning homework (Stigler and Hiebert

1999). The findings showed that the teachers documented in LPS showed little evidence of a consistent lesson pattern, but instead appeared to vary the structure of their lessons purposefully across a topic sequence. The analysis then focused on "lesson events," a viable unit for comparison, characterized by a combination of form (visual features and social participants) and function, such as intention, action, inferred meaning, and outcome (Clarke et al. 2006a, b). For example, Kikan-Shido (also known as between-desk instruction or seatwork) had a recognizable structural form evident across all classrooms. It was during the enactment of Kikan-Shido that each teacher's relative prioritization of monitoring or guiding student activity was most evident. Taking the documented Shanghai lessons as an example, the teacher in the lessons circulated and commented on the students' work. In some cases, he might be correcting errors, and in some cases he might be encouraging students to think further than their original solution (Lopez-Real et al. 2004). For the German lessons, the teacher's use of questioning to stimulate student mathematical thought during Kikan-Shido was very evident in the data (Clarke et al. 2006a). The analysis of Japanese lessons (Hino 2006) showed that following the seatwork activity, the teacher used the students' work for a variety of purposes, including: eliciting their mistakes, eliciting their puzzlement, eliciting opposing solutions, pointing out different solutions and giving explanations, pointing out difficulties and giving explanations, taking up students' questions and making their way of thinking visible to the group. The findings from this LPS study suggested additional reasons to those identified in the TIMSS Video Study about why the enactment of Japanese lessons differed from the other countries. The students' seatwork in Japanese lessons served the function of student exchange of information and opinions, and thinking about the problem together (Mok 2015).

2.3.3 Multiple Accounts of a Teacher's Practice

Another advantage of the LPS data set is to allow researchers to reconstruct multiple accounts of the classroom scenarios by putting together the data from all the lesson materials, including videos, student interviews, and teacher interviews, hence providing the opportunity for an in-depth study of the practice of a particular teacher in a specific cultural system. For the purpose of seeking an explanation for the "Asian Learner's Paradox,"[3] Mok (2006) analyzed 15 consecutive Grade 8 lessons by a Shanghai teacher. Based on the analysis of the students' interviews, the study provided evidence of students' active thinking in class and also a consistent concern for learning the mathematical content. Mok (2006) also argued that despite the strong teacher guidance in the lesson, the teacher showed an interpretation of student-centeredness that was different from the idea of student-centeredness in the

[3]The "Asian Learner's Paradox" refers to the apparently contradictory phenomenon of outstanding student performances in Asian regions and the reported classroom environment being non-conducive to learning with characteristics of directive teaching, large classes, etc. (Watkins and Biggs 2001).

Western education community. The teacher saw himself as non-traditional and he had made use of his deep understanding of his students in order to create a lesson experience so that the students might follow his intended plan with little side-tracking. The analysis of the sampled lessons showed that the teacher was very skillful in creating a "framed exploratory experience" for the topic of equations, rich in "variation" (Mok 2006). Variation has been featured as pertinent for learning to take place in the tradition of phenomenography and also seen as a characteristic of effective Chinese teaching of mathematics (Gu et al. 2004). An analysis of an episode is recapitulated here to show how the teacher attempted to guide the class to experience different levels of representations of a train-ticket problem.

2.3.4 The Episode of the Train-Ticket Problem

The problem is as follows:

> Xiu-min and his family went to Beijing for a holiday. They booked 3 adult tickets and 1 student ticket, costing a total of 560 dollars. His classmate Xiu-wang, learning this, decided to join Xiu-ming's family for the trip. Consequently, they bought 3 adult tickets and 2 student tickets, costing a total of 640 dollars. Please calculate the cost of 1 adult ticket and the cost of 1 student ticket

The problem was first solved mentally by a student, Dora, who was invited to share her solution to the class orally.

> Dora: Um, three adult tickets and one student ticket, um…totally five hundred and sixty dollars, and then adding one more student ticket will become six hundred and forty dollars. Deducting five hundred and sixty dollars from six hundred and forty dollars will be the price for one student ticket.

Dora's answer was arithmetic and intuitive in nature. The teacher immediately endorsed Dora's answer, but also repeated the answers with an emphasis on the idea of subtraction with the more technical phrase "the difference between …"

> T: The difference between the two is the price for one student ticket, which means the price for a student ticket should be six hundred forty dollars minus five hundred and sixty dollars, which is eighty dollars.

Following this, the teacher then asked the class to do the problem again using the method of equations. Such a request for a different way to solve the problem directed the students' attention to the method instead of the answers (the cost of the student and adult tickets). The teacher guided the students to write the equations $3x + y = 560$ and $3x + 2y = 640$, and then obtain the answer by subtracting one equation from the other. Comparing it with Dora's method, the teacher told the class explicitly that the method of equations was equivalent to Dora's method.

In this segment of the lesson, the teacher had created two levels of contrasts. The first level is the contrast between Dora's and the teacher's endorsed description of the student's method, giving a subtle emphasis of the method of subtraction. The second level was the contrast between the arithmetic method and the equation method for solving the same problem. At this second level, the students may need the teacher's help to see that the two apparently different methods are fundamentally the same method in this case. The teacher's intervention here guarantees a better chance for the students to see the contrast. However, how an individual student may learn depends on his/her own comprehension. After introducing these two levels of contrasts, the teacher discussed this new method of subtracting equations in parallel with the old method, "elimination-by-substitution," which they had learned in an earlier lesson. In this case, the students have to compare the new instance with what they are asked to recall. This is more abstract, therefore this a level 3 contrast.

The lesson presented in Mok (2006) is by no means spontaneous, but rather represents a synthesis based on the effort of a very experienced teacher and his understanding of a pedagogical framework of variation that is well recognized and implemented in Qingpu, his local region. Qingpu is a county of Shanghai (Experimenting Group of Teaching Reform in Mathematics in Qingpu County, Shanghai 1991) that has advocated creating an appropriate path for practicing the thinking process based on the variation of concepts, background and complexity of situations, and avoiding mechanical repetition in the design of exercises for practice. The strong teacher guidance in the lesson arose from the teacher's interpretation of student-centeredness, which was different from that in the Western education community. In the teacher's interview, he made a clear differentiation between his own teaching and traditional learning by rote. The teacher saw himself as non-traditional and he had made use of his strong understanding of his students in order to create a lesson experience that followed his intended plan (Mok 2006).

The conceptions of the teacher and the performance in the lesson were quite consistent with the findings of another study that compared conceptions of effective teaching between Chinese and U.S. teachers (Cai and Wang 2010). This suggested that the constraints of content coverage, teaching pace, and large class size affected teaching flexibility and student-centeredness. As far as the students' perspective is concerned, their interviews showed evidence of students' active thinking in class and also a consistent concern for learning the mathematical content. In addition, all students indicated that they liked the lesson and the teacher. While the teacher's teaching could be influenced by other factors such as examinations and societal expectations, these other factors might also have affected the students such that they welcomed their teacher's style of providing them with a framed exploratory experience in the classroom (Mok 2006). The analysis of the lessons of this Shanghai teacher on the one hand helps understanding the Asian Learner's Paradox in the context of the cultural background. On the other hand, it supports the argument that teaching is a cultural activity. Furthermore, in terms of methodology it gives an example of presenting an analytic account of effective teaching from multiple perspectives.

2.3.5 Lessons for the Implementation of Mathematical Tasks

Both the TIMSS Video Study and LPS have selected mathematical problems or tasks to be the focus in the comparison. The TIMSS Video Study compared the problems used in the lessons. Two key results for the mathematics lessons were reported:

(1) Teachers in all countries studied spent the majority of lesson time solving problems;
(2) Teachers in higher-achieving countries implemented making connections problems differently from teachers in the USA (Stigler and Hiebert 2004).

The analysis of the study considered two types of problems: "using procedure" problems, that is, problems requiring students to use only a memorized procedure or algorithm, and "making connections" problems, that is, problems requiring students to establish relationships between ideas, facts, and procedures and to engage in mathematical reasoning. With the exception of Japan, all six countries used more "using procedure" problems than "making connections" problems. Hence, the results show that the USA was no different from the higher-achieving countries in the kinds of problems that teachers *presented to students*. Then, what or where was the difference? As far as seeing teaching as a cultural activity was concerned, the videos of each country revealed some unique features. For example, Netherlands lessons frequently used calculators and real-world problem scenarios. The Japanese did neither in the sample, but the Japanese students spent on average a longer time working to develop their own solution procedures for problems that they had not seen before. Nonetheless, both countries had high levels of student achievement. Key result 2 above showed that how the teachers implemented "making connections" problems mattered. In all the high-performing countries except Australia the teachers made use of the rich potential in the problem statements and did not simplify the problem, i.e., they implemented a higher percentage of "making connections" problems as "making connections" problems. In contrast, the U.S. teachers changed "making connections" problems to "using procedure" problems, hence lowering the cognitive demand of the problems. In other words, the U.S. teachers let their students do problems of lower cognitive demand than the high performance region (Roth and Givvin 2008; Stigler and Hiebert 2004). Furthermore, the TIMSS Video Study showed that to improve the quality of teaching, the focus should be on the teaching and student experience rather than the superficial aspect of teaching, such as the organization, tools, curriculum, and textbooks. The way that the teacher and students interacted about the content could be more powerful than the curriculum materials the teacher used. The analysis and documentation of classroom practice provided a knowledge base for the teaching profession, enabling teachers to learn about teaching practice in other countries (Stigler and Hiebert 2004).

As far as comparison between different countries is concerned, the international team members of LPS have made some significant achievements in the use of mathematical tasks in classroom instruction (Shimizu et al. 2010). For example, Huang and Cai (2010) compared the implementation of mathematical tasks between the U.S. and Chinese lessons and were able to identify two levels of cognitive demands with four categories, namely, memorization, procedure without connection, procedure with connection, and doing mathematics. In the findings by Huang and Cai (2010), the sampled teachers of the LPS data for both USA and China were willing to present cognitively demanding tasks in their lessons and implement them by soliciting students' answers and organizing exploratory activities, yet the Chinese teacher was more frequently able to sustain the cognitive demand of the mathematical tasks during implementation. Mesiti and Clarke (2010) made a functional analysis of the mathematical tasks in the LPS data from China, Japan, and Sweden. They chose a novel method of selecting "distinctive" tasks, which might mean either typical or unusual; however, in both cases, the tasks were chosen for suggesting instructional effectiveness. How did they make this choice? In the researchers' interpretive reflection on the distinctive tasks, some important pedagogical values underpinning the choice were demonstrated by a few examples quoted here:

> The Stairs Task is well-known and used by mathematics teachers in many countries. The task was chosen for analysis because of its visual appeal and capacity to provide a focus for student attention;
>
> We chose this task (the Numbers task) in part because of its function in introducing students to the underlying structure of algebraic representations and in assisting them to develop appropriate mathematical language;
>
> The Graph Task was chosen for its strategy of 'creating' a real-world context for an abstract mathematical representation. The students became responsible for giving meaning to two lines on a graph and subsequently for determining the relationship in algebraic terms by identifying the value of the gradient;
>
> The Coin Task was chosen for its visual appeal, for the attempt at modeling the problem with a manipulative and for its non-routine nature;
>
> The Compare and Contrast task was chosen because of the open-ended nature of some aspects of the task (that is, to list similarities and differences) and because the sharing of students' work was essential for a full discussion to take place.

Mesti and Clarke (2010) further argued that the classroom performance of a task was ultimately a unique synthesis of task, teacher, students, and situation.

2.3.6 Lessons for Changing Classroom Instruction

International studies of classroom instruction very often are implicitly or explicitly associated with issues of effectiveness, the existence of national patterns or features, and what to compare and how to compare. Some of these issues have been addressed in the aforementioned studies with collaborative international efforts and novel inventions of methodology.

1. The TIMSS Video Study, building upon the tradition of large-scale survey of a national sample, suggested seeing teaching as a cultural activity and presented national patterns of reviewing, demonstration of the problem for the day, practicing

and correcting seatwork, and assigning homework. LPS, designed to complement other international studies such as the TIMSS Video Study, provided comparisons of mathematics lessons via analysis of lesson events during a sequence of lessons and multiple accounts, including the perspectives of the teacher and the learners.

2. Lessons events such as Kikan-Shido are lesson components found across different cultural systems. Although teachers in different cultural systems spend time on the same lesson event, they might be in fact carrying out activities with different meanings and functions. For example, the function of Kikan-Shido or between-desk instruction differed between teachers of different cultural systems. With the exception of the Japanese teachers, not all teachers make a strong link between the learning outcome of Kikan-Shido and the whole-class interaction episode. These lesson events were important components in classroom instruction with pedagogical functions and features, the feasibility of which may be further explored in other cultural systems.

3. Attempting to explain the Asian Learner's Paradox, the classrooms of Asian countries with a track record of better student performance, such as the Chinese classrooms, were sometimes the focus of investigation and comparison. The purpose of these studies is not to present a model for others to follow but rather to unfold an account of the enactment of mathematical tasks, through in-depth investigation of an effective case taking into account the many constraints (such as examination-orientation, content coverage, teaching pace, and large class size in a specific cultural system) and culturally rooted clues (such as the teacher's conceptions and beliefs, the norm of the students' expectation, the locally implemented pedagogical framework).

4. Seeking a common language for comparison has a specific implication for understanding effective instruction in different cultures. Both the TIMSS Video Study and the Learner's Perspective Study have chosen tasks as a unit for comparing and contrasting. Based on the quantitative comparison of "using procedure" problems and "making connections" problems, the TIMSS Video Study indicated that there were unique features in the teaching and student experiences despite the similar representation of problems. In addition to the nature of tasks or curriculum materials, how the teachers and students implement the tasks is very important. Although sometimes a task may have objectives in alignment with the objectives of promoting mathematical processes and making connections, in some cases teachers turn it into the routine of promoting rote learning. Therefore, the teachers' enactment of the teaching materials in alignment with the curriculum objectives is an important element of effective teaching (Stigler and Hiebert 2004). In addition, how the teacher sustains the intended roles of the tasks during implementation is also important (Cai 2005; Huang and Cai 2010).

5. Different kinds of tasks play different roles in the agenda of effective classroom instruction, as they are like the ingredients for preparing a meal—yet how the ingredients are cooked is important. The tasks employed in a lesson represent a model of mathematics as it is performed in that lesson (Mesiti and Clarke 2010). Among the many issues to be explored for effective classroom instruction, the

pedagogical models for different kinds of mathematical tasks catering to different values in different cultural systems will be a valuable agenda.

2.4 *Lesson 4: Making Global Research Locally Meaningful: TIMSS in South Africa*

Findings from both small-scale and large-scale studies have implications for improving learning and teaching. The findings from international comparative studies can also have policy implications. In this section, we discuss local policy implications of using TIMSS data in South Africa. We illuminate how a country can find its own voice in using the findings from the international comparative studies with a view to extend analyses which are meaningful to the local agenda.

Numerical and mathematical skills are globally recognized as key competences for the development of an individual, a society and an economy. The apartheid social engineering project in South Africa withheld mathematics as a school subject from the African population. The then Minister of Education pronounced in Parliament:

> …what is the use of teaching the Bantu[4] mathematics when he cannot use it in practice? The idea is absurd (Hansard 1953).

The result of these prejudices was an education policy that promoted racially differentiated access to mathematics within the framework of "Bantu education", which was designed to under-develop and exclude Black people.

Since 1994, the democratic government has emphasized the centrality of education, numeracy and mathematics. Performance in school mathematics is one of the indicators of the health of our educational system, and we recognize that under-performance continues to be a contributor to social inequalities of access and income. Further, the changes in school mathematics performance provide a measure of the extent of transformation since the inception of the democratic state. Within this context, the South African government, business and society have invested enormous resources to improve the quality of the education system, especially in mathematics.

As part of the strategy to improve education, government has included assessments and assessment studies to estimate the national performance in the key subject areas of mathematics, reading and science. Since 1994, the country has participated in a number of national, as well as cross-national comparative assessment studies to measure performance and understand the dynamics related to achieving higher educational quality. South Africa participated in TIMSS[5] in 1995 and subsequently in 1999, 2002, 2011 and 2015, and also participated in the

---

[4]Black South Africans were at times officially called "Bantu" by the apartheid regime.

[5]Testing mathematics and science proficiency at the Grade 8 and 9 levels.

regional Southern African Consortium for Monitoring Educational Quality (SACMEQ)[6] in 2000, 2007, and 2015.

Responses to the results from TIMSS have been mixed, with critics arguing that participation in international assessments is a pointless exercise because of the slow pace of improvement in South African education, and because these assessments are not relevant to the country's social and political history. Supporters have argued that international assessments (especially those with trend measures) can be useful at many levels of policy and planning. The goal of my participation in this panel is to share the experiences of the use of the TIMSS achievement datasets and information in South Africa to inform educational policy and extend the knowledge boundaries. This paper will be framed by three themes (i) Understanding South African mathematics achievement from 1995 to 2011, beyond the rank order; (ii) the contextual dynamics influencing mathematics achievement and (iii) extending the TIMSS sample to a 4 year South African Youth Panel Study (SAYPS) to understand educational progression and pathways. This paper will draw from the studies summarized in Table 2.

2.4.1 Mathematics Achievement Trend Over 20 Years

In 1994, the country had to reverse the separate and unequal policy of the apartheid government (set up along racial lines with a differential set of resources) to create a single, unified education department. Up to that point, there was no single national score estimating educational performance. Participation in TIMSS 1995 (with its methodological limitations and results not widely distributed) provided the first indicative estimate of national mathematics and science achievement. This was followed by the widely publicized results for TIMSS 1999 which lamented the low South African mathematics and science achievement scores, and the rank order which placed South Africa last of the set of 38 countries who participated. This international comparative study catalysed the debate about educational performance in South Africa, and involved many sectors of the society—politicians, policy-makers, academics, school teachers and the public. Newspaper headlines in South Africa asserted, for example, on the publication of mathematics scores in international studies (TIMSS 1999) that 'South African pupils are the dunces of Africa' (Sunday Times, 16 June 2000), 'Bottom of the class in maths', (Sunday Times, 14 October 2001), and 'Grade 3 flunkers sound a warning about our schools' (Sunday Times, 22 June 2003). The low mathematics performance and country rank order of last position repeated itself in the TIMSS 2003. The newspaper headlines and reaction from politicians and policymakers echoed those in the post TIMSS 1999 period, but the challenge for research was to embark on deeper analysis and extend the story from simply being the lowest performer to one which could provide policy directions.

[6]Testing mathematics and literacy at the Grade 6 level.

Table 2 Summary of the findings from analysis of TIMSS data

Area of analysis	Key findings
South African mathematics achievement	
Mathematics and science performance in grade 8 in South Africa 1998/1999	Low national mathematics mean score and last position on the rank order table
Mathematics and science achievement in South African Schools in TIMSS 2003	Low national mathematics mean score and last position on the rank order table. High educational inequalities reflective of the societal inequalities
Beyond Benchmarks: What 20 years of TIMSS data tells us about South African education	Low national mathematics mean score. High, but slightly reduced educational inequalities from 2002 to 2011. Trend analysis from 1995 to 2011 shows an improvement in mathematics achievement by 63 TIMSS points, equivalent to an improvement of 1.5 grade levels
Contextual factors influencing educational achievement	
Home SES indicators	
The presence of assets in the home promotes a learning environment Parental education levels have strong links with student performance	Positive as the number of assets increases Positive as levels of parental education increase
School SES Indicators	
Students in schools (using pre-1994 racial categories) that are better resourced and taught high quality curricula achieve better performances Independent and public fee paying schools with higher levels of infrastructure and resources than public no-fee paying schools have better learning environments	Positive as historical resource provision rises Higher resourced schools outperform the no-fee paying schools
Speaking the language of instruction at home	
Students who speak the language of instruction at home more frequently, tend to perform better than those who do not	Positive for convergence of languages, negative for divergence
Age	
Student achievement is higher for age-grade appropriate learners	Negative for learners who are older or younger than the expected age
Gender	
A difference in performance between male and female students indicates cultural and historical influences	The gender achievement gap is small to non-existent. Gender differences favoured girls for higher educational expectations, higher levels of parental engagement and experiencing lower levels of bullying
School safety	
Students in schools that are safer and where discipline is high tend to perform better	Positive in small to medium schools with fewer discipline or safety problems

(continued)

Table 2 (continued)

Area of analysis	Key findings
Student progression and pathways through secondary school	
Foundational mathematics skills	
Mathematics achievement gaps persist through secondary school	Mathematics performance in early grades is strongly predictive of survival to grade 12
Educational pathways and progression through secondary school	
TIMSS mathematics performance in grade 9 predicts educational pathways and performance in subsequent years	There is the predictable story of who succeeds in school, but there are also some students who succeed against the odds

Howie (2001, 2003), Reddy et al. (2006), Winnaar et al. (2011), Reddy et al. (2012), (2015a, b, c), Taylor et al. (2015), Visser et al. (2015), Isdale et al. (2016)

An important, but overlooked finding from the TIMSS analysis was the range of performance between the 5th and 95th percentiles of performance. Of all the countries participating in TIMSS 2003, South Africa had the widest range of scores between the 5th and 95th percentile. This reflects the wide disparities in society and in schools, and is evident in the educational outcomes of the students. The wide variation of the South African mathematics scores led to the characterization that there were two systems of education in the country, and the performance scores in TIMSS were reflective of South African inequalities (HSRC media briefing 2003).

The story of South African performance cannot told through the single national score, but through appropriate disaggregation. The disaggregation of the achievement scores by school type revealed that there was a strong correlation between socio-economic status and achievement scores. Africans, those who were most disadvantaged by the apartheid policies, had the lowest performance. African schools are located in areas where most Africans live and these areas are characterized by high levels of poverty and unemployment.

South Africa's participation in TIMSS 2011 provided an opportunity to measure the changes in educational performance since 1995. TIMSS was the only study that provided a scientifically rigorous methodology (in the country) to measure trends over the last 20 years (see Table 3). Analysis of the four rounds of TIMSS participation showed that the average national mathematics scale score remained the same over the years 1995, 1999 and 2002 (Reddy et al. 2012). In contrast, from 2002 to 2011 the national average mathematics score increased by 63 TIMSS points. The increase over the last two cycles of TIMSS can be translated to mean that overall student performance, though still low, has improved by one and a half grade levels. In 2011, the variance in the range of mathematics scores decreased, suggesting that the country is progressing (albeit slowly) towards more equitable educational outcomes.

The lack of change in the mathematics scores from 1995 to 2002 may be attributable to structural and educational changes as the country moved from apartheid to a democratic state. One of the unintended consequences of the many

Table 3 Trends in mathematics achievement for TIMSS 1995, 1999, 2002, and 2011

MATHS	Ave scale score	Score distribution
Grade 9 TIMSS 2011	352 (2.5)	229-516 = 287
Grade 9 TIMSS 2002	285 (4.2)	152-472 = 320
Grade 8 TIMSS 2002	264 (5.5)	117-484 = 367
Grade 8 TIMSS 1999	275 (6.8)	113-485 = 372
Grade 8 TIMSS 1995	276 (6.7)	142-496 = 354

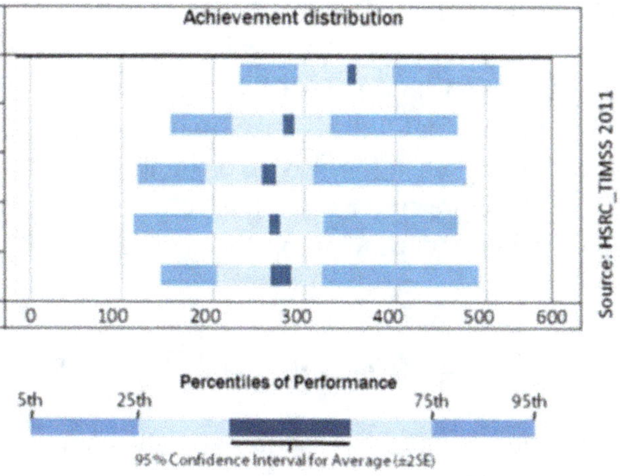

changes that occurred in the educational landscape was that school and classroom educational quality suffered, leading to continued poor educational outcomes.

2.4.2 Contextual Factors Influencing Educational Achievement

We need to move beyond the estimates of national mathematics achievement scores to investigate the factors that influence mathematical performance. The emerging results from our analyses confirm the effects of home and school socio-economic factors on mathematics achievement. As expected, students who speak the language of the test at home are more likely to achieve higher scores than those who do not. Similarly, the analysis confirms that student achievement is higher for age appropriate students.

We explored the effects of two contemporary, South African issues on achievement, namely gender and school violence. We found new complexities in the schooling experience of South African boys and girls. On average, across South Africa, gender differences in mathematics test results were small or non-existent. We also probed students about their attitudes to mathematics and found that both boys and girls valued mathematics. A particularly worrisome finding is the level of indifference among boys about their education. Boys were found to have lower aspirations about their academic careers, showed less interest in mathematics and engaged less often with an adult regarding their school work. The link between negative attitudes and weak performance was stronger for boys than for girls. Boys were also at a higher risk of being victims of bullying than girls.

The second issue we explored was the extent of violence in South African schools and the effects on mathematics achievements. Although concerns about school safety are increasing internationally, violence in schools is considered more serious in South Africa than elsewhere. We found that school related violence may be linked to conditions in communities. The extent of school safety largely depends on the type of school that learners attend. We found that children attending public schools experienced more frequent threats of violence than children attending independent schools. The socio-economic status of students is an indicator for potential exposure to acts of violence, with the chances of being bullied regularly being higher for students from poor families. There is a higher frequency of bullying for boys than for girls who attend schools with similar characteristics. Schools where there are fewer discipline or safety problems achieve better results, but this relationship is dependent on the size of the school.

2.4.3 Student Progression and Pathways Through Secondary School

In addition to concerns about low mathematics achievement, there is also a concern about how students progress through secondary schools. We analysed the pathways and performances in mathematics of secondary school students in South Africa using a panel-like data set of Grade 8 students who participated in the TIMSS 2003 and who were tracked in the Grade 12 examination data sets (Reddy et al. 2012). Firstly, we found that students who started with similar Grade 8 mathematics scores had different educational outcomes 4 years later. Secondly, in middle class schools, Grade 8 mathematics scores were a good indicator of who would pass the exit level examination, whilst this relationship was not as strong in schools for poorer students. Thirdly, there was a stronger association between TIMSS Grade 8 mathematics scores and subject choice of secondary school mathematics in middle class schools than in poorer schools. Fourthly, there was a strong correlation between Grade 8 mathematics performance and the Grade 12 examination mathematics achievement. Mathematics performance in the earlier years predicted later mathematics performance. This study adds to the body of evidence that shows that to improve educational outcomes, the policy priority should be to build foundational knowledge and skills in numeracy.

To extend our understanding about the pathways and transitions followed by South African youth, the first wave of the South African Youth Panel Study (SAYPS) was administered in 2011. SAYPS, a longitudinal panel study, followed grade 9 learners who participated in TIMSS 2011, 4 years, to explore the educational transitions of young people. We found that students followed one of four educational pathways (Table 4) through secondary school.

Table 4 Educational pathways of students in the South African Youth Panel Study

Smooth	Staggered	Stuck	Stopped
Neat, year-on-year grade progression through school	Learners in school for all four waves of SAYPS, who make some grade progress but have at least one episode of grade repetition	Learners in school for all four waves of SAYPS, but stuck in grade 9 or 10 for three or more periods	Individuals who leave school before Wave 4 and do not return

Almost half the sample, at 47 %, followed the smooth pathway, while 39 % followed a staggered pathway and 14 % were either stuck or stopped. There is a predictable story of 'advantage begetting advantage' for students who experience a smooth pathway: these students tend to have a higher than average TIMSS score, parents with a higher level of education, positive attitudes about school and come from homes with more books. Our analyses show that there is also a new story, and it is possible to succeed academically despite disadvantage. Almost 57 % of those who followed a smooth pathway came from middle class schools, and just over 43 % of this group came from non-fee paying schools for poorer students. We will study this group further to understand their pathways to success.

There is an emerging scholarship using TIMSS data to understand South African mathematics performance and contribute to the debates for improved educational performance. We recognize the limitations of cross-national comparative studies, but given our limited resources (financial and human), it is more prudent that the TIMSS data and opportunity is used optimally to extend knowledge, inform policy, and provide strategic intelligence to inform the educational directions in the country. The research challenges for the use of the TIMSS data are multiple: measuring performance, tracking changes over time, and establishing the key policy-amendable determinants for improved achievement scores.

3 Summary and Looking Ahead

Over the last three decades, international comparative studies have completely transformed the way we see mathematics education. For example, because of the very high ranking of some Asian countries, the field of mathematics education has become interested in mathematics education in Asian countries. We used to think that there was one basic way of teaching mathematics; international comparative studies, however, showed us many different ways of teaching mathematics in the classroom. We also learned that some student background variables (e.g., attitudes, gender) operate in different ways for students in different countries.

In this Topical Survey, we have reflected on a sample of the lessons we can learn from international comparative studies. To conclude, we would like to emphasize the complementary roles of small-scale and large-scale international comparative studies. While we have used 'small' and 'large' scale throughout this paper, we have not described what we mean about the scale. The small- and large-scale international comparative studies can address different research questions. To make this point clear, we cite the report from the Institute for Education Sciences and National Science Foundation in the USA (U.S. Department of Education & National Science Foundation 2013), which presented six classes of research in their *Common Guidelines for Education Research and Development*: (1) Foundational Research, (2) Early Stage or Exploratory Research, (3) Design and Development Research, (4) Efficacy, (5) Effectiveness, and (6) Scale-up. From investigating new phenomena in Exploratory research, or development of new theories in

Foundational research, to the examination of the effectiveness of interventions across a wide range of demographic, economic, and implementation factors in Scale-Up research, scale is a critical factor in establishing the believability of our conceptual models and the potential efficacy of our designed innovations in mathematics education. Small-scale studies tend to be used towards the beginning of a research program to explore new phenomena. Large-scale studies, in contrast, tend to be employed after such methods or instruments have been piloted and their use justified, and the phenomena to which they apply have been adequately defined. Anderson and Postlethwaite (2007) define the purpose of large-scale studies as describing a system as a whole and the role of parts within it. But the complexity of the system, and the type of understanding to be gained from the study, greatly impact how large the scale must be.

There are a number of advantages of large-scale international comparative studies, such as their utility for understanding situations and trends, the technical capacity they provide for testing hypotheses involving a number of variables, and the sophisticated analytic methods often employed in such studies (Cai et al. 2015; Middleton et al. 2015). The international comparative studies, such as TIMSS and PISA report a wide range of achievement variables, school variables, system variables, and student variables, and there are literally thousands of individual results for each survey. As a multitude of commentaries also attest, the implications of any of these results for educational policy must be derived with careful attention to local situations and to the details of the measurement instruments. There are also resource and time limitations for conducting large-scale international comparative studies. Because of less resource and time constraints, small-scale international comparative studies, on the other hand, can offset some of the pitfalls of large-scale international comparative studies by providing valuable analysis that is relevant to the situation of the particular participants. Small-scale studies allow for in-depth analysis of issues being studied. Thus, while small-scale and large-scale international comparative studies are conducted separately, their findings and analysis could complement to each other for us to understanding and improving students' learning.

An alternative complementary role of the small-scale and large-scale studies is to conduct in-depth analyses for some issues based on the large-scale data. For example, go back to PISA's three processes of Formulate, Employ, and Interpret. When we compare students from Australia, China, Singapore, and the USA on the three processes, we found a different pattern. For Australia and USA students have relatively better performance on interpreting than on the other two processes, while Shanghai and Singapore students did a better job on formulating than on the other two processes. With in-depth analysis from large-scale data, we can find such subtle, but important differences, but we need a range of other studies to understand why this might be the case (Table 5).

In addition to small-scale and large-scale studies, LPS gives an example of another possible kind of international collaboration that serves a complementary role to the other international studies and provides opportunities for studying the artifacts of classroom instruction by reconstruction of the classroom data from

Table 5 Students' mean scores (and standard deviations) in PISA's three processes

	Employ	Formulate	Interpret	Overall
Australia	500.46 (94.88)	497.83 (110.25)	514.15 (101.04)	504.15 (96.29)
Shanghai-China	612.76 (92.86)	624.42 (119.37)	578.74 (97.60)	612.68 (100.98)
Singapore	574.09 (97.72)	581.69 (122.06)	555.13 (105.52)	573.47 (105.36)
United States	480.11 (90.55)	475.52 (98.08)	489.56 (95.84)	481.37 (89.96)

multiple perspectives, thereby enhancing the understanding of the possible pedagogical functions and values in different cultural systems.

Finally, we would like to point out the fact that mainstream mathematics education research journals publish very few articles based on the large-scale international comparative studies. The results and background theories from the large-scale international comparative studies are usually published in books and public media and specialist journals (e.g. on assessment and scaling). This is despite the fact that the data is generally publically available for analysis by mathematics educators. In contrast, mathematics education research journals publish a lot more small-scale international comparative studies. Large-scale studies are resource-intensive—both in terms of financial and human resources—and it is important to use the results for more than the measurement of performance and rank order position to develop a scholarship around these resource intensive studies. As for individual researchers, because of less resource and time constraints, engaging in small-scale international comparative studies could be much more fruitful to future endeavor.

References

Alexander, R. (2000). *Culture and pedagogy: International comparisons in primary education*. Oxford: Blackwell Publishing.

Anderson, L. W., & Postlethwaite, T. N. (2007). *Program evaluation: Large-scale and small-scale studies*. International Academy of Education: International Institute for Education Planning. http://www.unesco.org/iiep/PDF/Edpol8.pdf

Asian Society. (2006). *Math and science education in a global age: What the U.S. can learn from China*. New York, NY: Asian Society.

Bharucha, J. (2008). America can teach Asia a lot about science, technology, and math. *The Chronicle of Higher Education, 54*(20), A33.

Bradburn, M. B., & Gilford, D. M. (1990). *A framework and principles for international comparative studies in education*. Washington, DC: National Academic Press.

Cai, J. (1995). A cognitive analysis of U.S. and Chinese students' mathematical performance on tasks involving computation, simple problem solving, and complex problem solving. *Journal for Research in Mathematics Education Monograph Series 7*. Reston, VA: National Council of Teachers of Mathematics.

Cai, J. (2000). Mathematical thinking involved in U.S. and Chinese students' solving process-constrained and process-open problems. *Mathematical Thinking and Learning, 2*, 309–340.

Cai, J. (2004). Why do U.S. and Chinese students think differently in mathematical problem solving? Exploring the impact of early algebra learning and teachers' beliefs. *Journal of Mathematical Behavior, 23*, 135–167.

Cai, J. (2005). U.S. and Chinese teachers' knowing, evaluating, and constructing representations in mathematics instruction. *Mathematical Thinking and Learning, 7*(2), 135–169.

Cai, J., & Ding, M. (2015). On mathematical understanding: Perspectives of experienced Chinese mathematics teachers. *Journal of Mathematics Teacher Education,*. doi:10.1007/s10857-015-9325-8.

Cai, J., & Silver, E. A. (1995). Solution processes and interpretations of solutions in solving a division-with-remainder story problem: Do Chinese and U.S. students have similar difficulties? *Journal for Research in Mathematics Education, 26*(5), 491–497.

Cai, J., & Wang, T. (2010). Conceptions of effective mathematics teaching within a cultural context: Perspectives of teachers from China and the United States. *Journal of Mathematics Teacher Education, 13*(3), 265–287.

Cai, J., Ding, M., & Wang, T. (2014). How do exemplary Chinese and U.S. mathematics teachers view instructional coherence? *Educational Studies in Mathematics, 85*(2), 265–280.

Cai, J., Hwang, S., & Middleton, J. A. (2015). The role of large-scale studies in mathematics education. In J. A. Middleton, J. Cai, & S. Hwang (Eds.), *large-scale studies in mathematics education* (pp. 405–416). New York, NY: Springer.

Clarke, D., Emanuelsson, J., Jablonka, E., & Mok, I. A. C. (Eds.). (2006a). *Making connections: Comparing mathematics classrooms around the world*. Rotterdam: Sense Publishers B.V.

Clarke, D., Keitel, C., & Shimizu, Y. (Eds.). (2006b). *Mathematics classrooms in 12 countries: The insiders' perspective*. Rotterdam: Sense Publishers B.V.

Elley, W. B. (2002). *Evaluating the impact of TIMSS-R (1999) in low- and middle-income countries*. http://www.iea.nl/fileadmin/user_upload/Publications/Electronic_versions/Elley_Impact_TIMSS-R.pdf

Experimenting Group of Teaching Reform in maths in Qingpu County, Shanghai. (1991). *Xuehui jiaoxue [Learning to teach]*. Beijing, China: People Education Publishers.

Friedman, T. L. (2005). *The world is flat: A brief history of the twenty-first century*. Farrar, Straus, and Giroux.

Gu, L., Huang, R., & Marton, F. (2004). Teaching with variation: A Chinese way of promoting effective mathematics learning. In L. Fan, N.-Y. Wong, J. Cai, & S. Li (Eds.), *How Chinese learn mathematics: Perspectives from insiders*. Singapore: World Scientific Publishing Co.

Hansard. (1953). Debates of the South African Parliament.

Hino, K. (2006). The role of seatwork in three Japanese classrooms. In D. Clarke, C. Keitel, & Y. Shimizu (Eds.), *Mathematics classrooms in 12 countries: The insiders' perspective* (pp. 59–74). Rotterdam: Sense Publishers B.V.

Howie, S. J. (2001). *Mathematics and science performance in grade 8 in South Africa. 1998/1999*. Pretoria, Human Sciences Research Council.

Howie, S. J. (2003). Language and other background factors affecting secondary pupils' performance in Mathematics in South Africa. *African Journal of Research in Mathematics, Science and Technology Education, 7*(1), 1–20.

Huang, R., & Cai, J. (2010). Implementing mathematical tasks in US and Chinese Classrooms. In Y. Shimizu, B. Kaur, R. Huang, & D. Clarke (Eds.), *Mathematical tasks in classrooms around the world* (pp. 145–164). Rotterdam: Sense Publishers.

Isdale, K., Reddy, V., Winnaar, L., & Zuze, L. (2016). *Smooth, Staggered or Stopped: Educational Transitions in the South African Panel Study*. Human Sciences Research Council: Labour Market Intelligence Partnership Publication.

Juan, A., Zuze, T. L., Reddy, V., Namome, C., & Hannan, S. (2015). *Does it matter if students enjoy learning science? Exploring student attitudes towards science in South Africa.* http://www.timss-sa.org.za/?wpfb_dl=41

Lloyd, G. M., Cai, J., & Tarr, J. E. (2016). Research issues in curriculum studies: Evidence-based insights and future directions. In J. Cai (Ed.), *Compendium for Research in Mathematics Education*. National Council of Teachers of Mathematics: Reston, VA.

Lopez-Real, F. J., Mok, I. A. C., Leung, F. K. S., & Marton, F. (2004). Identifying a pattern of teaching: An analysis of a Shanghai teacher's lessons. In L. Fan, N.-Y. Wong, J. Cai, & S. Li (Eds.), *How Chinese learn mathematics: Perspectives from insiders* (pp. 382–412). Singapore: World Scientific Publishing Co.

Ma, L. (1999). *Knowing and teaching elementary mathematics: Teachers' understanding of fundamental mathematics in China and the United States*. Hillsdale, NJ: Erlbaum.

Mesiti, C., & Clarke, D. J. (2010). A functional analysis of mathematical tasks in China, Japan, Sweden, Australia and the USA: Voice and Agency. In Y. Shimizu, B. Kaur, R. Huang, & D. J. Clarke (Eds.), *Mathematical tasks in classrooms around the world* (pp. 185–216). Rotterdam: Sense Publishers.

Middleton, J. A., Cai, J., & Hwang, S. (2015). Why mathematics education needs large-scale research. In J. A. Middleton, J. Cai, & S. Hwang (Eds.), *Large-scale studies in mathematics education* (pp. 1–16). New York, NY: Springer.

Mok, I. A. C. (2006). Shedding light on the East Asian Learner Paradox: Reconstructing student-centredness in a Shanghai classroom. *Asia Pacific Journal of Education, 26*(2), 131–142.

Mok, I. A. C. (2015). Research on mathematics classroom practice: An international perspective. In J. C. Sung (Ed.), *Selected regular lectures from the 12th International Congress on Mathematical Education* (pp. 589–606). New York: Springer.

Organisation for Economic Co-operation and Development (OECD). (2013a). *PISA 2012 assessment and analytical framework: Mathematics, reading, science, problem solving and financial literacy*. Paris: OECD Publishing.

Organisation for Economic Co-operation and Development (OECD). (2013b). *PISA 2012 results: What students know and can do (Volume I)*. PISA: OECD Publishing.

Organisation for Economic Co-operation and Development (OECD). (2013c). *PISA 2012 results: Ready to learn students' engagement, drive and self-beliefs (Volume III)*. PISA: OECD Publishing.

Postlethwaite, T. N. (1988). Preface. In Postlethwaite, T. N. (Ed.), *Encyclopedia of comparative education and national systems of education* (pp. xvii–xxvi). Oxford: Pergamon.

Reddy, V., Kanjee, A., Diedericks, G., & Winnaar, L. (2006). *Mathematics and science achievement in SA schools in TIMSS 2003*. Cape Town: HSRC Press.

Reddy, V., Van der Berg, S., Janse van Rensburg, D., & Taylor, S. (2012). Educational outcomes: Pathways and performance in South African high schools. *South African Journal of Science, 108*(3/4).

Reddy, V., Zuze, T., Visser, M., Winnaar, L., & Juan, A. (2015a). *Have we researched gender equity in Mathematics education?*. Policy brief, Cape Town: HSRC Press.

Reddy, V., Zuze, T., Visser, M., Winnaar, L., Juan, A., & Hannan, S. (2015b). *Safe and sound? Violence and South African education*. Policy brief, Cape Town: HSRC Press.

Reddy, V., Zuze, T., Visser, M., Winnaar, L., Juan, A., Prinsloo, C., et al. (2015c). *Beyond benchmarks: What twenty years of TIMSS data tells us about South African education*. Cape Town: HSRC Press.

Roth, K., & Givvin, K. B. (2008). Implications for math and science instruction from the TIMSS 1999 Video Study. *Principal Leadership, 8*(9), 22–27.

Shimizu, Y., Kaur, B., Huang, R., & Clarke, D. (Eds.). (2010). *Mathematical tasks in classrooms around the world*. Rotterdam: Sense Publishers.

Silver, E. A., Leung, S. S., & Cai, J. (1995). Generating multiple solutions for a problem: A comparison of the responses of U.S. and Japanese students. *Educational Studies in Mathematics, 28*(1), 35–54.

Song, M. J., & Ginsburg, H. P. (1987). The development of informal and formal mathematical thinking in Korean and U.S. children. *Child Development, 58*, 1286–1296.

Stacey, K., & Turner, R. (2015a). *Assessing mathematical literacy: The PISA experience*. Heidelberg: Springer.

Stacey, K., & Turner, R. (2015b). PISA's reporting of mathematical processes. In K. Beswick, T. Muir, & J. Wells (Eds.), *Proceedings of 39th Psychology of Mathematics Education Conference* (Vol. 4, pp. 201–208). Hobart, Australia: PME.

Stevenson, H. W., Lee, S., Chen, C., Lummis, M., Stigler, J. W., Liu, F., et al. (1990). Mathematics achievement of children in China and the United States. *Child Development, 61*, 1053–1066.

Stigler, J. W., & Hiebert, J. (1999). "The teaching gap." *Best ideas from the world's teachers for improving education in the classroom*. New York: The Free Press.

Stigler, J. W., & Hiebert, J. (2004). Improving mathematics teaching. *Educational Leadership, 61* (5), 12–17.

Stigler, J. W., Lee, S., & Stevenson, H. W. (1990). *Mathematical knowledge of Japanese, Chinese, and American elementary school children*. Reston, VA: NCTM.

Stigler, J. W., Gallimore, R., & Hiebert, J. (2000). Using video surveys to compare classrooms and teaching across cultures: Examples and lessons from the TIMSS video studies. *Educational Psychologist, 35*(2), 87–100.

Taylor, S., van der Berg, S., Reddy, V., & Janse van Rensburg, D. (2015). The evolution of educational inequalities through secondary school: Evidence from a South African panel study. *Development Southern Africa, 32*(4), 425–442.

U.S. Department of Education, & National Science Foundation. (2013). *Common guidelines for education research and development*. http://www.nsf.gov/publications/pub_summ.jsp?ods_key=nsf13126

Visser, M., Juan, A., & Feza, N. (2015). Home and school resources as predictors of mathematics performance in South Africa. *South African Journal of Education, 35*(1), 1–10.

Watkins, D. A., & Biggs, J. B. (Eds.). (2001). *Teaching the Chinese learner*. Hong Kong: Comparative Education Research Centre, The University of Hong Kong.

Winnaar, L. D. F. G., Blignaut, R., & Frempong, G. (2011). Understanding school effects in South Africa using multilevel analysis: Findings from TIMSS 2011. *Electronic Journal of Research in Educational Psychology, 13*, 151–170.

Zhao, Y. (2008). What knowledge has the most worth? Reconsidering how to cultivate skills in U. S. students to meet the demands of global citizenry. *The School Administrator, 74*(4), 48–52. http://www.aasa.org/publications/saarticledetail.cfm?ItemNumber=9737

www.ingramcontent.com/pod-product-compliance
Lightning Source LLC
LaVergne TN
LVHW060059220926
848484LV00008BA/2385
* 9 7 8 1 0 1 3 2 6 7 5 7 4 *